Separations Division 2024

Held at the 2024 AIChE Spring Meeting and 20th Global Congress on Process Safety

New Orleans, Louisiana, USA
24-28 March 2024

ISBN: 979-8-3313-1703-4

Printed from e-media with permission by:

Curran Associates, Inc.
57 Morehouse Lane
Red Hook, NY 12571

Some format issues inherent in the e-media version may also appear in this print version.

AIChE
120 Wall Street, FL 23
New York, NY 10005-4020

Phone: (800) 242-4363
Fax: (203) 775-5177

www.aiche.org

Additional copies of this publication are available from:

Curran Associates, Inc.
57 Morehouse Lane
Red Hook, NY 12571 USA
Phone: 845-758-0400
Fax: 845-758-2633
Email: curran@proceedings.com
Web: www.proceedings.com

TABLE OF CONTENTS

58a A Quantum Chemical Study on the CO_2-Amine Reaction Pathway

Nahoko Kuroki, Amol Patil, Hirotoshi Mori

To achieve the targets set in the Paris Agreement, global CO_2 emissions must be reduced by 7.6% annually over the next decade. This requires an effective use of carbon capture and storage (CCS) technologies. However, one of the barriers to the use of such systems is their absorption phase. Carbon capture costs are primarily influenced by the energy required for solvent regeneration, directly related to CO_2–solvent interactions and the separation efficiency from other gases. Improving the efficiency lowers the carbon capture cost, facilitating the use of CCS technologies.

With this background, developing efficient and sustainable CO_2-absorbing liquids has received immense attention in chemical engineering and related fields. While there are several types of CO_2 absorbers, in our previous works, we have intensively studied physisorbents, that is, ionic liquids (ILs; room-temperature molten salts [1]) and deep eutectic solvents (DESs; hydrogen-bonded mixtures in a broad sense [2]). Assisted by statistical thermodynamic calculations and machine learning results, a new phosphonium-based ionic liquid was experimentally proven to have higher CO_2 solubility than the previous best. However, the high viscosity of ILs prevented their practical application.

This study discusses the solvation effect (dielectric constant of the solvent) on the amine-CO_2 chemical reaction from a quantum chemical viewpoint to develop new amines that chemisorb CO_2 with low energy cost.

[1] N. Kuroki, Y. Suzuki, D. Kodama, F. A. Chowdhury, H. Yamada, H. Mori, Machine Learning-Boosted Design of Ionic Liquids for CO_2 Absorption and Experimental Verification, *J. Phys. Chem. B* **127**, 9, 2022–2027 (2023).

[2] S. Watabe, N. Kuroki, H. Mori, COSMO-RS Exploration of Highly CO_2-Selective Hydrogen-Bonded Binary Liquid Absorbents under Humid Conditions: Role of Trace Ionic Species, *ACS Omega* **8**, 16, 14478–14483 (2023).

58b A Systematic Study on the Influence of Oriented Nanoparticles in Polymer Nanocomposite

Abayomi Adeleke, Pedro Arce, Robby Sanders

Native gel electrophoresis is a powerful technique for the separation of proteins based on their size and shape that also enables subsequent assessments of protein function (such as through Western blotting). Such assessments are important in many areas including in bio-health applications such as medical diagnostics. In this study, we examined the impact of bisacrylamide concentration and bentonite nanoparticles on the native gel electrophoresis of two model proteins, ovalbumin and carbonic anhydrase. The experiments involve varying the concentration of bisacrylamide in the presence and absence of bentonite nanoparticles (a mineral ubiquitous in clay) in the gel matrix. The research is relevant to fostering understanding of the role of the hydrogel structure on the separation performance of proteins. For example, we hypothesize that the results would reveal that increasing the concentration of bisacrylamide leads to a decrease in the mobility of both proteins. Furthermore, the addition of bentonite should result in a more pronounced retardation effect on ovalbumin when compared to the relatively smaller carbonic anhydrase. These findings would be a first step in understanding the impact of crosslinker and the presence of additives in polyacrylamide gel electrophoresis (PAGE) and can significantly impact the separation efficacy and characterization of proteins by native gel electrophoresis. Further investigations would attempt to understand the underlying mechanisms of these effects and their potential applications in drug delivery and related developments. Preliminary results and trends will be reported in this contribution.

58c Design of Heat-Integrated Pressure Swing Distillation Process Using a Graphical Pinch Analysis

Donggun Kim, Chaeyeong Seo, Minyong Lee, Jae Lee, Heecheon Lee

Based on a graphical pinch analysis, a method for the fundamental design and heat integration of pressure swing distillation (PSD) is demonstrated. The pinch pressure, determined by the graphical pinch analysis, serves as a minimum pressure constraint that enables the provision of the required temperature driving force for heat integration and bypass of distillation boundaries. To evaluate the feasibility of heat-integrated pressure swing distillation (HIPSD), the proposed method necessitates the visualization of feasible composition regions for initial feed compositions and pinch temperature contour lines, considering pressure changes and the dynamic characteristics of a three-component mixture. Therefore, in this study, an in-depth investigation was conducted through three case studies to demonstrate heat integration in the HIPSD. The HIPSD system can have different pinch pressures depending on the distillation column arrangement or the separation sequence of products. Furthermore, considering pinch constraints, the proposed analysis method allows for the design of energy-efficient HIPSD processes by determining the optimal operating pressure of the columns.

58d Prediction of Main Transition Velocities and Gas Holdups in a Bubble Column Operated with Aqueous Alcoholic Solutions

Stoyan Nedeltchev, Sara Marchini, Markus Schubert, Uwe Hampel

Bubble columns (BCs) are the most important gas-liquid contactors. They are frequently used in the chemical industry (for absorption, oxidation, chlorination, etc.) because these reactors are capable of providing excellent mass and heat transfer characteristics. Depending on their specific application, BCs could operate either in the homogeneous flow regime (FR) (mainly used in biotechnological applications) or heterogeneous FR (used in 80 % of the industrial applications). In between, the transition FR exists, which is characterized with single coalesced bubbles but without the presence of liquid circulation. That is why two transition velocities U_{trans}, distinguishing the boundaries of the three main FRs, are usually identified. The first one is associated with the onset of bubble coalescence, whereas the second one distinguishes the onset of liquid macro-circulation.

The operation of BCs with foaming systems has not been exhaustively investigated. In this work, a new reliable FR identification approach as well as reasonable prediction of gas holdups in the homogeneous and transitions FRs will be demonstrated for conditions with low foaming.

Gauge pressure fluctuations have been recorded in a BC (0.1 m in ID) equipped with a perforated plate gas distributor (96 orifices × Ø 1.0 mm). The BC operated with mixtures of deionized water (DW) and 2-pentanol with different volume concentrations (0.5, 1.0, 1.5 and 2.0 vol. %). The signal consisted of 10,000 points, which were subdivided into 10 intervals (each consisting of 1000 points). The local average absolute deviation (AAD_i) in each interval was estimated together with the mean AAD (AAD_{mean}) from all ten divisions. Based on this approach, the average relative error (ARE) in each interval was calculated. It was found that the dimensionless ratio of maximum ARE to total ARE (sum of all ten ARE) is capable of identifying reliably the main transition velocities. In other words, a modified error analysis was performed. For instance, in case of DW + 1.0 vol. % 2-pentanol three transition velocities U_{trans} (at 0.044, 0.055 and 0.077 m/s) were identified. The third U_{trans} value distinguished the onset of the complete foaming state. This way, a successful FR identification can be performed by means of only one statistical parameter (AAD is a robust statistical parameter about the data width around the mean of the signal) and error analysis. It is noteworthy that the first U_{trans} value practically coincides with the prediction by the empirical correlation of Im et al. (2019). In a similar way, the three U_{trans} values were successfully identified at the other volume concentrations of 2-pentanol in DW (0.5, 1.5 and 2.0 vol. %).

The empirical fitting (average relative error < 5 %) of 40 experimental overall gas holdups to the superficial gas velocity in the homogeneous and transitional FRs of the BC (0.1 m in ID) operated with the same gas-liquid mixtures was compared with the predictions of the empirical correlation of Sotello et al. (1994). The result shows that for foaming systems the gas holdups can be predicted by previously developed correlations in non-foaming systems, however, a small modification should be applied.

References

Im, H., J. Park and J. W. Lee, *ACS Omega* **4**, 1329-1343 (2019)

Sotello, J. L., F. J. Benitez, J. Beltran-Heredia, C. Rodriguez, *Int. Chem. Eng.* **34**, 82-91 (1994)

58e A Novel Helical Descending Liquid-Bridge for Continuous Demulsification

Liang Yuan, Haifeng Cong, Xingang Li

Existing physical demulsification methods are expensive and offer low separation efficiencies. Therefore, the development of efficient and green continuous demulsification processes has garnered research attention. Inspired by the novel helical liquid bridge flow realized previously by our group, a continuous demulsification method featuring a helically descending liquid bridge was established in this study to treat oil-in-water emulsions with low oil contents. First, the formation of the liquid bridge flow was verified. Then, the continuous demulsification of the helically descending liquid bridge was monitored using a high-speed camera, which helped elucidate the demulsification mechanism. Furthermore, the surface of the helical string was modified to make it oleophilic for improving the demulsification efficiency. The results showed that the helical liquid descent achieved with surface modification led to a competitive separation efficiency of up to 52.94%, which was 27.94% higher than that of other helical microchannel devices with hydrophobic surfaces.

46a Troubleshooting Distillation Columns

Andy Burleigh

In all chemical plants and refineries there generally is not enough instrumentation in the right places in our processes. During the design phase, decisions were made between knowing more about the process, which means buying more instruments, and saving capital funds. When a problem appears with the operation of a vessel, there are always gaps in the knowledge of what is happening due to a lack of instruments in the places where knowledge is not routinely required. Process engineers and operations personnel need to know what is happening so they can take corrective action. When knowledge is missing, they often take actions that make matters worse. For distillation and separation processes Chemical Engineers have historically relied upon plant process measurements such as low rates, temperatures, and pressures and model or simulation results for data with which to solve troubleshooting or re-design problems. An array of on-line diagnostic services are available that can provide additional data offering real-time information on how pieces of process equipment are actually operating. The most common application is Gamma Scanning. This test is primarily applied, but not necessarily limited to, distillation or separation columns. Gamma scans provide a density profle of the internal process of operating distillation columns and other process vessels. The density profle can be used to diagnose the hydraulic operating conditions of mass transfer devices such as damage to internals, fooding, degree of entrainment or weeping, liquid levels on trays and distributors, liquid distribution through packed beds, etc. The presenter will show case studies where scanning revealed vital process information that helped solve an operating problem or helped make a revamp/re-design successful. Additional derivatives from gamma scanning of detecting flow distribution patterns will be demonstrated. The ThruVision Scan technique is a specialized horizontal gamma scan used to generate a topographic profile of the internal cross-sectional density of process equipment. This profile is useful for the detailed study of liquid flow distribution through packed columns as well as having applications beyond distillation columns. A hybrid application involving tracers and scanning on fixed bed catalytic reactors will also be discussed.

46b Ideal Stage Vs. Rate-Based Simulation of Gas Treating Processes

G. Simon A. Weiland

Rate-based simulation for sour gas treating applications is a relatively new, but proven, technology that was developed in the 1980s and 1990s. The older more traditional technology uses the concept of the ideal and has been used to simulate acid gas removal applications throughout the industry for many years, so what are the advantages to using rate-based simulators? The discussion will cover the conceptual differences and technical bases between mass transfer rate-based simulators and ideal-stage based simulators, why someone would choose one over the other, and the advantages and disadvantages of each.

26a Practical Industrial Distillation

Daniel Summers

This presentation, aimed at young professionals in Separations, provides an overview of industrial distillation - from theory to practice.

26b Distillation Column Hydraulics

Anand Vennavelli

Distillation column hydraulics encompass a range of phenomena. These include the vapor and liquid interactions, various flow patterns, changes in pressure, entrainment, weeping, vibrations, liquid distribution, and several flooding mechanisms, among others. This paper offers a concise overview of these crucial hydraulic events in distillation columns and explores their measurement and modeling methods.

26c Evolution of Distillation Equipment Technology and Future Development

Tony Cai

Distillation is the most important unit operation in chemical process industry. Both trayed and packed columns are widely used in distillation and absorption applications. Over past 100 years, many new generation distillation trays and packings that exhibit better hydraulic and mass transfer performance had been developed. In this presentation, the evolution of the tray and packing technologies in past 10 years will be reviewed. The latest innovation and development on distillation equipment will be discussed. The areas for future research and development on tray and packing technologies and other column internals will be presented.

AUTHOR INDEX